To Auntie Marion and Uncle Winston,
who gave me books—keys that unlocked
so many doors.

—L. A.

THE CURIOUS LIFE OF

CECILIA PAYNE

DISCOVERING THE STUFF OF STARS

WRITTEN BY

LAURA ALARY

ILLUSTRATED BY

YAS IMAMURA

EERDMANS BOOKS FOR YOUNG READERS

GRAND RAPIDS, MICHIGAN

When she was eight years old, Cecilia Payne discovered she was a scientist.

She was lying in the orchard under the apple trees when she spied a flower that was not supposed to be there. It looked like a purple star with a bee nestled at its heart. Although she had never seen one before, she knew it by name: the bee orchid. Cecilia had heard her mother talk about seeing these flowers in the south of France. She had seen drawings of them in books. This blossom nodding in the tall grass perfectly matched the picture in her mind.

BEE ORCHID

Her mother said it was impossible. The bee orchid does not grow in England.

But Cecilia knew what she had seen. And she was right.

Cecilia carefully dug up the bee orchid and moved it to her garden. Every day she visited the little plant and made a promise to herself. She would learn to see all the wonders of nature. And she would always trust what she knew was right, no matter who said it was impossible.

As it turned out, Cecilia was good at seeing—
especially hidden things.

Other children saw pretty pebbles and filled their pockets. But Cecilia saw fragments of ancient mountains, tumbled smooth by time and a journey as old as the Earth.

Other children looked up at the stars and imagined fantastic things: celestial campfires, twinkling diamonds, holes snipped in the velvet dome of heaven. But Cecilia saw a mystery to be solved. She wanted to know what the stars *really* were.

Cecilia was always exploring. She got to know the trees up close, spent time face-to-face with flowers, called the constellations by name. At school she learned to look carefully, measure accurately, and store facts in her memory. She wanted to read everything—even the encyclopedia! Her teacher told her she could learn to read anything she wanted—poetry, music, ancient Greek, the language of numbers. So she did.

When she was twelve years old, Cecilia moved to London. She was homesick for the countryside, the green fields, the wildflowers and birds. At her new school, Cecilia felt like the bee orchid—out of place.

No one understood why she wanted to collect plants and draw pictures of them. No one understood why she wanted to spend hours with test tubes and chemicals. No one understood why, when she won an award at school, she chose a book about fungi as her prize.

But Cecilia knew what she wanted.

More than anything, she wanted to go to university. When she won a scholarship to the University of Cambridge, it seemed that a whole new world was opening up.

Those were exciting days at Cambridge. Famous scientists were seeing what no one could see with their eyes, hidden things that were visible only to the imagination.

The insides of atoms.
Vast and distant stars.
The mind-bogglingly small.
The infinitely large.

Cecilia wanted to see these things, too.

One night Cecilia went to hear a talk by Arthur Eddington, the famous astronomer in charge of the Cambridge Observatory.

His words lit a fire inside her. That very night she decided she wanted to be an astronomer.

Cecilia found an old telescope at the college where she lived. She repaired it and began to explore the sky for herself.

She saw wonderful things. The moons of Jupiter. Saturn's rings. Cecilia filled notebooks with ideas, observations, and questions.

When she visited the Cambridge Observatory, Cecilia asked so many questions that the flustered guide went scurrying to Arthur Eddington for help. The great astronomer suggested a few books, but Cecilia had already read every single one.

So Eddington invited her to use the observatory library. It was as if he had given her a key that unlocked the doors of the heavens. Now she could read for herself the thoughts of the greatest astronomers, past and present.

Cecilia wanted to be one of them.

Yet even at Cambridge, Cecilia was out of place. In her classes she was often the only woman.

Some professors made her sit by herself. Others ignored her. They said that she would never be an astronomer. That women do not belong in the world of stars and atoms.

So Cecilia packed her bags and crossed the ocean to a place where there was room for her.

She came to Harvard, where she tasted a new kind of freedom. She had her own rooms and could stay up all night reading if she chose.

Best of all, she was surrounded by women studying the stars! These women had gotten to know the stars the way you get to know people. Slowly. Patiently. By spending time with them and paying attention.

They gathered around Cecilia and shared with her their greatest treasure—thousands of photographs of starlight. Hidden in those glass photographic plates was a mystery to be solved.

These women knew the colors of the stars. Their brightness. Their positions in the sky. But they did not know what the stars were made of. Neither did Cecilia. Yet. But she brought with her from Cambridge fresh ideas that would help her find out.

When starlight shines through a prism, it is broken apart and the rainbow of colors—the spectrum—hidden inside can be seen. On each of those glass photographic plates was a spectrum, and on each spectrum was a pattern of black lines. What did they mean? Starting with what her eyes could see, Cecilia let her mind carry her further.

Cecilia knew that different kinds of matter absorb different colors of light. When light from the heart of a star passes through its atmosphere, some of that light is soaked up by elements in the atmosphere, leaving gaps in the spectrum. Could those black lines be the key? Could they tell her what stars are made of? Using what she had learned at Cambridge, Cecilia found a way to read the patterns.

In those days, astronomers believed stars were made of the same stuff as Earth—just much hotter. After studying the photographs and patterns in the starlight, Cecilia reached a different conclusion: stars were made mostly of gases—hydrogen and helium.

Night after night Cecilia sat up late and checked her work. When she felt sure, she wrote down her conclusions and how she reached them. But it is hard to be the only one who sees something.

Other astronomers said it was impossible.
That Cecilia must be mistaken.

Cecilia knew what she had seen in the photographs and the numbers. But was it real? For the first time, Cecilia doubted herself. Could she really be right—and all those great astronomers be wrong?

Long ago Cecilia had promised herself she would always trust what she knew was right. For the first time, she broke that promise and backed down.

But what she had seen *was* real. The stars are *not* like Earth. They are blazing balls of gas.

It took those other astronomers a few years to catch up.
But at last they came to see what Cecilia saw.

When she became professor of astronomy at Harvard University, Cecilia taught her students to look carefully, measure accurately, and store facts in their memories. She welcomed their questions. She taught them to trust themselves and what they knew was right—no matter who said it was impossible. She had learned that lesson the hard way.

Becoming a scientist, she told her students, is not an easy path. You see things no one else gets to see. Sometimes people doubt you. Sometimes you doubt yourself. It can be a tough climb up a steep hill. But if you keep going, the view is amazing.

As usual, Cecilia was right.

AUTHOR'S NOTE

Twinkle, twinkle little star.
How I wonder what you are!

Most children know this song. But how many know the woman who was the first to figure out what the stars *really* are? I had never heard of Cecilia Payne until an episode of *Cosmos* introduced me to this determined young woman—a pioneer in the field of astrophysics—whose research led her to challenge the views of the most senior astronomers of her day. I wanted to know more about her! The more I learned about Cecilia Payne, the better I liked her.

HOW CURIOUS!

Cecilia's curiosity was boundless. She was interested in *everything*. At various times in her life she played the violin, conducted orchestras, dabbled in woodworking, took to the stage (she once played the title role in *Richard III*), tried her hand at bookbinding, and made up stories for her younger brother and sister. She learned to read many languages, including German, French, Italian, Russian, Latin, ancient Greek, and Icelandic. Above all, she loved mathematics and the sciences, especially chemistry, which let her peer beneath the surface of the material world and see what the fabric of the universe was really made of.

When Cecilia had a question, she chased the answer—through books, by seeking out people who could help her, or by getting her hands dirty and figuring it out for herself. No matter what the obstacle, she found a way around it. Cecilia was left-handed, and in those days teachers forced all students to write with their right hands. Cecilia found this painful, so she researched the problem, and taught herself to write with both hands—and also upside-down, backward, and upside-down-backward!

STANDING OUT OR FITTING IN?

Perhaps it is not surprising that Cecilia had trouble fitting in. She stood out because she was different: exceptionally tall, often dressed in unfashionable hand-me-down clothes, keenly intelligent, awkward at small talk, and fascinated by subjects most people couldn't care less about. But although she was sometimes lonely, Cecilia never tried to change or disguise herself.

She may have been shy in social settings, but when Cecilia set her mind on something, she was bold about going after it. For instance, when her professors told her there was no future for her as an astronomer in England, she confronted Harlow Shapley, head of the Harvard College Observatory, who was visiting London, and told him she wanted to come to America and work for him. Cecilia persisted until Shapley offered her not only a place at Harvard, but a scholarship too. The scholarship was not quite enough, so Cecilia wrote an award-winning Greek essay and used the prize money to pay for new clothes for her trip.

KNOW YOURSELF AND TRUST YOURSELF

Cecilia's resourcefulness was one aspect of the unique kind of self-confidence she possessed. Some people seem confident because they proclaim their opinions loudly. But real confidence does not need to be loud. It does not need to perform or seek attention. Cecilia was confident because she worked hard and knew how to do thorough and careful research. When she was sure she was right about something, she stuck to her position because she trusted herself.

Only once did this confidence falter. When a famous astronomer told her she could not possibly be correct about the stars being made of hydrogen and helium, Cecilia doubted herself. But this makes me like her all the more—because we all have weak moments. What matters is what we do next. Get angry? Call ourselves failures? Blame other people? Cecilia did none of these. She just carried on with her work until it became clear that she had been right all along. Then she used her story as a lesson for her students, as a reminder that even the most eminent scholars can be wrong—so do your work carefully, trust yourself, and keep an open mind.

WHERE ARE YOU IN THIS STORY?

I wanted to tell the story of Cecilia Payne so that younger readers would have a chance to get to know her much sooner than I did—and maybe learn some important things from her. Scientist Vera Rubin once said that she became an astronomer partly because when she was a child, she read a book about Maria Mitchell and realized that women could do this work too, no matter who said otherwise! Seeing ourselves in stories can make a big difference.

Of course you don't have to be a girl to be inspired by Cecilia Payne. Maybe you just feel different. Out of place. Maybe you know what you want to be and do, but there seem to be too many obstacles blocking your way. Maybe you will think of Cecilia and take heart. I hope so.

— LAURA ALARY

CECILIA'S CURIOUS LIFE

1900 **May 10, 1900 – Cecilia Payne is born in Wendover, England.** Her father, Edward John Payne, is a London barrister, historian, and musician. Cecilia's mother, Emma Leonora Helena Pertz, is an artist. Cecilia is surrounded by music and books.

1904 **December 26, 1904 – Cecilia's father dies suddenly.** Her mother is left alone to raise Cecilia and her siblings. They often go to London to visit her aunts, whose circle includes many famous scientists. Cecilia grows up under the influence of strong, intelligent women.

1906 **1906 – Cecilia starts school at a small day school near her home.** Her teacher, Elizabeth Edwards, trains her students in memory, precise and accurate measurement, and careful observation. By age twelve, Cecilia has a good command of arithmetic, geometry, and algebra, and can read French, German, and Latin.

1908 **1908 – Cecilia correctly identifies a bee orchid in the orchard.** At this moment (one she later recalls as a "thunderclap of understanding") she first realizes that careful observation can help her understand the world.

1912 **1912 – Cecilia's family moves to London.** She is sent to St. Mary's College, a strict religious school with very little teaching in science. Cecilia is lonely and unhappy. She gets out of going to chapel by pretending to faint, and asks a local bookbinder to bind two works by Plato (a Greek philosopher) into one volume and emboss *Holy Bible* on the spine so her teachers will think she is studying scripture when she is really reading Greek philosophy! When she can, Cecilia creeps upstairs to a little room on the top floor used for upper-year science classes. There, she worships in her own way, "adoring the chemical elements."

1913 **1913 – Cecilia finally gets a proper science teacher.** Dorothy Dalglish arrives at the school to teach botany, chemistry, and atomic theory. She takes Cecilia under her wing. Sadly, Miss Dalglish falls ill in 1914 and stops teaching. The school does not replace her. Cecilia does her best to teach herself science and advanced mathematics. But the school does not give her the support she needs and eventually tells her to leave.

1917 **1917 – Cecilia begins studying at St Paul's Girls' School.** With its laboratories and science teachers, St Paul's is a dream come true for Cecilia. She shines in other subjects too. The music teacher, Gustav Holst, makes her part of the orchestra and encourages her to pursue a career in music. Cecilia is among the first to hear a new piece of music Holst is composing: *The Planets*.

1919 **1919 – Cecilia wins a scholarship to Cambridge.** She enrolls at Newnham College (the college for women) and begins to study botany.

December 2, 1919 – Cecilia hears a life-changing lecture by Arthur Stanley Eddington. Eddington has just returned from an expedition photographing stars during a solar eclipse to test Einstein's theory that gravity can bend light. Cecilia is so captivated by his account that she resolves to become an astronomer herself, switching her studies from botany to physics. Despite her enthusiasm, Cecilia is not welcomed. She does find some allies, though, including L. J. Comrie, who shows her how to fix and use the old telescope at Newnham Observatory, and teaches her the mathematics of computational astronomy.

1923 **1923 – Cecilia completes her studies at Cambridge University.** However, because at the time Cambridge will not grant degrees to women, she does not officially graduate. Realizing there is no opportunity for her to have a career as an astronomer in England, Cecilia looks elsewhere. At a lecture in London, she meets Harlow Shapley, director of the Harvard Observatory, and tells him she would like to come work for him.

Fall 1923 – Cecilia leaves England and sails to America. She is the first PhD student in astronomy at Radcliffe College (later part of Harvard University) and the Harvard College Observatory.

1924 **1924 – Cecilia publishes her first paper.** Initially, she is going to publish as C. H. Payne, but when Shapley asks if she is ashamed of being a woman, she decides to use her full name: Cecilia H. Payne.

1925 **1925 – Cecilia completes her PhD thesis: *Stellar Atmospheres: A Contribution to the Observational Study of High Temperature in the Reversing Layers of Stars.*** Its radical conclusions contradict the theories of famous astronomers like Eddington at Cambridge and Henry Norris Russell at Princeton University. When Russell reads Cecilia's thesis, he admires her work but feels certain she must be mistaken—though in 1929 he publishes his own paper admitting that Cecilia had been right.

1933 **1933 – Cecilia meets Sergei Gaposchkin.** Gaposchkin, a Russian astronomer working in Germany, fears for his life as the Nazis come to power. He spends four days bicycling to a conference in Göttingen in hopes of meeting Cecilia, whose scientific papers he has read. Sergei gives her a handwritten history of his life and begs her to help him get to the US. Cecilia promises to help, and by December of 1933, Sergei is working at Harvard Observatory.

1934 **March 1934 – Cecilia Payne and Sergei Gaposchkin are married.** They have three children: Edward (1935), Katherine (1937), and Peter (1940).

1938 **1938 – Cecilia is promoted to the rank of astronomer at Harvard.** However, she does not have the salary or job security of a professor. Although she is a brilliant teacher and her classes are popular, her name isn't even listed next to her classes in the course calendar until 1945.

1956 **1956 – Dr. Cecilia Payne-Gaposchkin is appointed Professor of Astronomy.** She is the first woman to become full professor at Harvard University.

1976 **1976 – Cecilia receives the Henry Norris Russell Prize.** Founded in 1946, this award from the American Astronomical Society celebrates lifetime achievement in the field of astronomy. Cecilia is the first woman to receive the honor.

1979 **December 7, 1979** – After a final trip around the world with Sergei, Cecilia is diagnosed with lung cancer and dies at home.

Photo: Cecilia at the Harvard College Observatory

REACHING FOR THE STARS

How do you study something that is too far away to see or touch? For a long time, this was the problem faced by anyone who was curious about the stars. All you could do was gaze from a distance.

Traditional astronomy was all about *observation* and *location*—knowing what was in the sky and where to look for it. Astronomers measured and mapped stars, calculated the movements of planets, and predicted the paths of comets. While some curious people may have wondered what the stars were really made of, this was not a question anyone could answer. No one could reach up and pluck a star from the sky to examine it!

But one step at a time, scientists found a way to use their minds to carry them from what they *could* see and touch, to what they could not.

Here are a few of the steps along the path that led to what Cecilia saw.

SECRETS IN THE STARLIGHT

1666 **1666 – Isaac Newton** shone ordinary sunlight through a glass prism and saw a rainbow of colors hidden inside the white light. He called this rainbow a *spectrum*.

1815 **1815 – Joseph von Fraunhofer** combined a glass prism with a small telescope so he could look at the spectrum in more detail. On the rainbow he could see patterns of dark lines. Fraunhofer had made the first *spectroscope*. But what were the lines?

1859 **1859 – Robert Bunsen** and **Gustav Kirchhoff** used the spectroscope to look at a different kind of light. When they heated various elements until they glowed, they found that each element made its own pattern of colored lines. These bright lines seemed to match some of the dark lines in the spectrum of the sun's light. The same elements glowing in the lab must be part of the sun too!

1872 **1872 – Henry Draper** was the first to take a photograph of the spectrum of a star. Those early images, captured on glass plates, were very tiny and hard to read—just small gray smears. But a patient and keen-eyed person could find the patterns hidden in the fine black lines.

1882 **1882 –** When Henry Draper died, his widow and collaborator **Mary Anna Draper** gave his collection of photographs to the Harvard Observatory to encourage further research. **Edward Pickering**, head of the observatory, wanted to use the glass plates to answer new questions: *What are stars made of? Why do they give off light? Why are some brighter than others?* But he needed help reading the plates and classifying the different types of stars. To carry out this important work, he hired women—an unusual choice in those days. These women became known as the **Harvard Computers**.

MYSTERIOUS MATTER: THE ATOM

Meanwhile, scientists in Europe were turning their minds to a new puzzle: *atoms*. Many believed that all matter was made up of basic particles called atoms (*atom* means *indivisible*). In the decade before Cecilia was born, scientists began to suspect that atoms were more complicated than anyone had imagined.

1897 **1897 – J. J. Thomson** (working at the Cavendish Laboratory at Cambridge where Cecilia would later be a student) demonstrated that all atoms contain tiny negatively charged particles. (We now call these *electrons*.) This was evidence that atoms had smaller parts.

1898 **1898 – Marie Curie** was studying some strange rays that had been discovered a few years earlier by Henri Becquerel. The rays turned out to be tiny particles being spat out of the atoms themselves! Curie called this phenomenon *radioactivity*.

1907 **1907 – Ernest Rutherford** used radioactivity to help reveal the inner structure of atoms. He aimed a stream of radioactive particles at a thin sheet of gold foil. Most of the particles passed through, but a few bounced back. Rutherford realized that the particles must have collided with something very small and dense inside the gold atoms. He called this the *nucleus*. Atoms, as Rutherford imagined them, were like miniature solar systems, with a nucleus at the center surrounded by a lot of empty space and some orbiting electrons.

1913 **1913 – Niels Bohr** came up with a different picture of the atom, saying electrons travel around the nucleus in fixed orbits, each with its own energy level. Electrons can jump from a lower to a higher orbit when they take in (absorb) energy, or from higher to lower when they give off (emit) energy. This absorption and emission only happens in specific amounts.

PUTTING THE PIECES TOGETHER

1920 **1920 – Meghnad Saha** was the first to connect atoms with the stars. He realized that those mysterious patterns of lines in starlight were caused by electrons jumping from one energy level to another.

If light from a star came to us directly, its spectrum would look somewhat like one continuous rainbow. But light has to pass through the outer atmosphere of the star, so some of its energy is soaked up by elements in the atmosphere. Because atoms of each element can only absorb energy in specific amounts, and these energies correspond to specific wavelengths of light, the presence of an element in the atmosphere creates a pattern of gaps (dark lines) in the spectrum where those wavelengths are missing.

1923 **1923 –** When **Cecilia Payne** left England for Massachusetts, she brought with her all these new ideas. But the ideas needed to be tested on real data—spectra from actual stars. The collection of photographic plates at Harvard Observatory (and the women who knew them best) gave Cecilia what she needed.

MORE ABOUT CECILIA'S WORLD

WHO WERE THE HARVARD COMPUTERS?

We think of computers as objects—electronic devices. But before the word developed that meaning, "computer" referred to a person who did mathematical calculations. At the Harvard Observatory, most of the computers were women.

Edward Pickering, head of the observatory in the late 1800s, needed help sorting and classifying the growing collection of photographs of stellar spectra (photographs of starlight passed through a prism and captured on thousands of glass plates). This work demanded mathematical ability, but also a lot of patience and precision.

One day Pickering got frustrated with one of his male assistants and declared that his maid could do a better job! His wife suggested that he should give his maid a chance. Williamina Fleming, who had been a schoolteacher in Scotland, turned out to be an excellent choice. She went on to develop a classification system used around the world, and became the first woman to hold a titled position at Harvard University: Curator of Astronomical Photographs.

Edward Pickering hired many other women as computers. Some, like Annie Jump Cannon and Henrietta Swan Leavitt, had college degrees in astronomy and physics. Another, Antonia Maury, had studied astronomy at Vassar College with Maria Mitchell—one of the first women to become a professor at an American university.

When Cecilia Payne arrived at the Harvard Observatory, she became part of this community of intelligent and educated women. What she brought that was new was her experience at the famous Cavendish Laboratory at Cambridge and her understanding of atomic physics.

WHAT CECILIA SAW . . .

When Cecilia Payne arrived at the Harvard Observatory in 1923, she was met by a complicated puzzle: hundreds of glass plates, with markings that looked like tiny gray smears patterned with vertical black lines. The plates were stellar spectra, created when starlight is gathered by a telescope, then fanned out by a prism to show the different wavelengths of light. These glass plates held a complicated code that Cecilia wanted to decipher—a code that just might help her discover what exactly the stars were made of.

Cecilia knew some things for certain. She knew that the black lines showed where energy from the starlight had been absorbed by elements inside the stars. She knew what some of those elements were, but not *how much* of each was present. She knew that the patterns in the spectra were connected to the temperature and pressure of the stars but was not sure precisely *how*.

At Cambridge, Cecilia had learned about the structure and behavior of atoms. She understood that at high temperatures, electrons (the negatively charged particles around the central nucleus of an atom) jump to higher energy states and are sometimes stripped away completely (a process called *ionization*). Building on the work of other scientists, she searched the spectra for signs that ionization was happening. After months of studying the plates without success, Cecilia had a breakthrough. She realized that the hotter the star, the more atoms were ionized—and she found proof of this in the spectra. The pieces of the puzzle began to slide into place.

What did Cecilia see? She saw that the patterns in the spectra showed what was happening inside stars, *within the atoms themselves*. She saw that though the spectra from each star might look very different, all the stars were primarily made of the same elements. The variations between spectra were caused by the ways that stars ionize elements differently depending on how hot they were. And she saw that, rather than being made up of the same heavy elements that make up our planet, the stars were mostly made of the universe's lightest elements: hydrogen and helium! This conclusion came as a shock to Cecilia—and everyone else. The idea that stars were giant balls of gas seemed impossible.

WHAT CECILIA HELPED OTHERS SEE . . .

But Cecilia was right. Thanks to her, we have learned to see the stars—and the rest of the universe—differently.

Cecilia Payne was the first person to discover what stars are made of. Her work showed astronomers what a powerful tool physics can be and helped lead to the growth of a new field of study: astrophysics. Her discovery of the abundance of hydrogen in stars gave rise to new questions—like *why* is there so much hydrogen in the universe? Looking for answers led scientists all the way back to the birth of the cosmos!

And as she continued to research, teach, and write books, Cecilia also helped other people see that women *do* belong among the stars and atoms.

HONORS AND AWARDS GIVEN TO CECILIA PAYNE

- Member of Royal Astronomical Society (1923)
- First recipient of the Annie J. Cannon Award in Astronomy (1934)
- Member of the American Philosophical Society (1936)
- Member of the American Academy of Arts and Sciences (1943)
- Award of Merit from Radcliffe College (1952)
- Rittenhouse Medal from the Rittenhouse Astronomical Society at the Franklin Institute (1961)
- Professor Emerita of Harvard University (1967)
- Asteroid 2039 Payne-Gaposchkin is named after Cecilia (1977)
- Honorary Degrees from Rutgers University, Wilson College, Smith College, Western College, Colby College, and the Women's Medical College of Pennsylvania.
- The Payne-Gaposchkin Patera (volcano) on Venus is named after Cecilia (1997)
- The American Physical Society's Doctoral Dissertation Award in Astrophysics is renamed the Cecilia Payne-Gaposchkin Doctoral Dissertation Award in Astrophysics (2018)

FURTHER READING

MORE ABOUT CECILIA PAYNE AND THE HARVARD COMPUTERS

Burleigh, Robert. *Look Up! Henrietta Leavitt, Pioneering Woman Astronomer*. Illus. Raúl Colón. New York: Simon & Schuster/Paula Wiseman Books, 2013.

Gerber, Carole. *Annie Jump Cannon, Astronomer*. Illus. Christina Wald. New Orleans: Pelican Books, 2011.

Larson, Kirsten W. *The Fire of Stars: The Life and Brilliance of the Woman Who Discovered What Stars Are Made Of*. Illus. Katherine Roy. San Francisco: Chronicle Books, 2023.

Lasky, Kathryn. *She Caught the Light: Williamina Stevens Fleming: Astronomer*. Illus. Julianna Swaney. New York: HarperCollins, 2021.

MORE ABOUT STARS AND ASTROPHYSICS

Kregenow, Julia. *Twinkle Twinkle Little Star, I Know Exactly What You Are*. Illus. Carmen Saldaña. Naperville, IL: Sourcebooks, 2018.

Tyson, Neil deGrasse, and Gregory Mone. *Astrophysics for Young People in a Hurry*. New York: Norton Young Readers, 2018.

SELECTED BIBLIOGRAPHY

Ghose, Shohini. *Her Space, Her Time: How Trailblazing Women Scientists Decoded the Hidden Universe.* Cambridge, MA: MIT Press, 2023.

Hirshfeld, Alan. *Starlight Detectives: How Astronomers, Inventors, and Eccentrics Discovered the Modern Universe.* New York: Bellevue Literary Press, 2014.

Koopman, Rebecca A., and A. G. Davis, eds. *The Starry Universe: The Cecilia Payne-Gaposchkin Centenary; Proceedings of a Symposium Held at the Harvard-Smithsonian Center for Astrophysics, Cambridge, Massachusetts, October 26–27, 2000.* Schenectady, NY: L. Davis Press, 2000.

Moore, Donovan. *What Stars Are Made Of: The Life of Cecilia Payne-Gaposchkin.* Cambridge, MA: Harvard University Press, 2020.

Payne-Gaposchkin, Cecilia. *Cecilia Payne-Gaposchkin: An Autobiography and Other Recollections.* Edited by Katherine Haramundanis. 2nd ed. Cambridge: Cambridge University Press, 1996.

Rubin, Vera C. "Cecilia Payne-Gaposchkin (1900-1979)." In *Out of the Shadows: Contributions of Twentieth-Century Women to Physics*, edited by Nina Byers and Gary Williams. Cambridge: Cambridge University Press, 2006.

Smith Zrull, L. "Women in Glass: Women at the Harvard Observatory during the Era of Astronomical Glass Plate Photography, 1875–1975." *Journal for the History of Astronomy* 52, no. 2 (May 2021): 115–46, DOI: 10.1177/00218286211000470.

Sobel, Dava. *The Glass Universe: How the Ladies of the Harvard Observatory Took the Measure of the Stars.* New York: Viking, 2016.

Published in 2025 byEerdmans Books for Young Readers, an imprint of Wm. B. Eerdmans Publishing Co. · Grand Rapids, Michigan · www.eerdmans.com/youngreaders

34 33 32 31 30 29 28 27 26 25 1 2 3 4 5 6 7 8 9

ISBN 978-0-8028-5515-2 · A catalog record of this book is available from the Library of Congress. · Illustrations created with gouache and watercolor

Eerdmans Books for Young Readers would like to thank Dr. Katie Mack, astrophysicist and Hawking Chair in Cosmology and Science Communication at The Perimeter Institute in Waterloo, Ontario, Canada, for sharing her expertise and reviewing this book for accuracy.

Image of Cecilia Payne-Gaposchkin courtesy of Smithsonian Institution Archives. Image # SIA2009-1325.